The Evolution of Ebola

by microbiologist and investigative author:

Gwendolyn Olmsted

First published by Amazon.com November 2014

© Copyright 2014 Gwendolyn Olmsted

The moral right of the author has been asserted.

This book is licensed for your personal enjoyment and learning only. This book may not be re-sold or given away to other people. No part of this book may be used or reproduced without written permission.

Book Class: Investigative Scientific Research

1st Edition: 17 November 2014

Available in paperback and *Kindle* ® e-book formats

Publisher: ***Amazon*** ®

© Copyright 2014 book cover design by Gwendolyn Olmsted

Table of Contents:

The Evolution of Ebola
Summary
Prologue

Chapter 1: Ebola virus—What is it?
Chapter 2: Initial Case and Reservoirs
Chapter 3: Virology
Chapter 4: Epidemiology and Outbreaks
Chapter 5: The Reston ebolavirus
Chapter 6: Research and Vaccines
Chapter 7: Initial 1976 Discovery
Chapter 8: The 2014 Strain
Chapter 9: Current Death Count and Outlook

Postscript
References
About the Author

Summary

Initially not discovered until 1976 in Sudan, Africa, the Ebola virus was named after the Ebola River near where its first outbreak occurred. Belgian doctor Peter Piot, a 27-year-old scientist and medical school graduate training as a clinical microbiologist, was sent to the region nearly forty years ago to find out why people were dying of a mysterious illness. He uncovered the Ebola virus, yet averted his attention to the virus of the hour, AIDS upon return, so his research abruptly stopped on Ebola. Now Director of the London School of Hygiene and Tropical Medicine, he discusses the initial 1976 strain of Ebola and how it has evolved into the 2014 strain it is today. Virology, epidemiology and outbreaks, Reston and Marburg virus comparisons, research and vaccines, and current death count and outlook are also examined in this handy survival guide to the 2014 Ebola Virus Disease (EBOV) pandemic.

Prologue

Human history becomes more and more a race between education and catastrophe

~H.G. Wells

Wait for the common sense of the morning.

~H.G. Wells

Chapter 1: Ebola virus—What is it?

Ebola virus disease (EBOV), or Ebola hemorrhagic fever, is a disease of humans and other mammals caused by the Ebola viruses classification of viruses. The incubation period the virus needs to take effect in humans is as early as two days from infection up to as long as three weeks. Signs and symptoms include a fever, sore throat, muscle pain, and headaches, at first. Then, transgresses into vomiting, diarrhea, rashes, and decreased function of liver and kidneys.

At this time those infected begin to bleed both internally and externally. All of their internal cells and organs begin to hemorrhage and rupture upon themselves, often leaving the infected crying blood and bleeding from their pores.

EBOV has a high risk of death, with those infected given a fifty percent chance of survival. This is most commonly because of low blood pressure due to rapid fluid loss.

Once EBOV has taken onset, the symptoms can last from one to three weeks.

EBOV spreads by direct contact with blood or other bodily fluids of an infected human or animal. Infection with the Ebola virus may also occur by direct contact with a recently contaminated item, including surfaces where the item may have been placed. Spread of EBOV through aerial transmission, through the air, has not been documented in natural, non-lab, environments. Ebola virus may also be spread by semen or breast milk, even for several weeks to months after recovery.

Fruit bats are believed to be the normal carrier in the natural environment, and are able to spread the virus without being affected or contracting EBOV themselves. However, humans become infected by contact with the bats or with a living or dead mammal that has been infected by bats. After human infection occurs, the disease may also spread between people.

Scientists have noted that other diseases such as malaria, cholera, typhoid fever, meningitis, and other viral hemorrhagic fevers may resemble EBOV. As a consequence, blood samples are testes for viral RNA, viral antibodies, and for the Ebola virus itself to confirm a diagnosis of EBOV.

Control of outbreaks requires coordinated medical services, along with a certain level of community engagement. The medical services include: Rapid detection of cases of disease; contact tracing of those who have come into contact with infected individuals;

quick access to laboratory services; proper care and management of those who are infected; and proper disposal of the dead through cremation.

Prevention includes limiting the spread of EBOV from infected animals to humans. This may be done by handling potentially infected bush meat only while wearing protective clothing and by thoroughly cooking it before consumption. It also includes wearing proper protective clothing and washing hands when around a person with the EBOV. Furthermore, samples of body fluids and tissues from people with the disease should be handled with special caution.

No vaccine for the Ebola virus is commercially available. Efforts to help those who are infected are supportive only; they include rehydration as well as treating specific symptoms. This supportive care improves outcomes.

EBOV was first identified in 1976 in an area of Sudan (now part of South Sudan), and in Zaire (now the Democratic Republic of the Congo). EBOV has thus far only occurred in outbreaks in the tropical regions of sub-Saharan Africa.

Through 2013, the World Health Organization (WHO) reported a total of 1,716 cases in 24 outbreaks. The largest outbreak to date is the ongoing epidemic in West Africa, which is centered in Guinea, Sierra Leone, and Liberia. As of November 2014, this outbreak has totaled 14,413 reported cases, resulting in 5,504 deaths, and counting.

Chapter 2: Initial Case and Reservoirs

Although it is not entirely clear how Ebola initially spread from animals to humans, the spread is believed to involve direct contact with an infected wild animal or fruit bat. Besides bats, other wild animals sometimes infected with EBOV include several monkey species, chimpanzees, gorillas, and baboons.

Animals may become infected when they eat fruit partially eaten by bats carrying the virus. Fruit production, animal behavior and other factors may trigger outbreaks among animal populations.

Evidence indicates that both domestic dogs and pigs can also be infected with EBOV. Dogs do not appear to develop symptoms when they carry the virus, and pigs appear to be able to transmit the virus to at least some primates. Although some dogs in an area in which a human outbreak occurred had antibodies to EBOV, it is unclear whether they played a role in spreading the disease to people.

The natural reservoir for Ebola has yet to be confirmed; however, bats are considered to be the most likely candidate species. Three types of fruit bats (*Hypsignathus monstrosus*, *Epomops franqueti* and *Myonycteris torquata*) were found to possibly carry the virus without getting sick. As of 2013, whether other animals are involved in its spread is not known. Plants, arthropods and birds have also been considered possible viral reservoirs.

Bats were known to roost in the cotton factory in which the first cases of the 1976 and 1979 outbreaks were observed, and they have also been implicated in Marburg virus infections in 1975 and 1980. The Marburg virus (MARV) very closely resembles the Ebola virus under the microscope as well as in symptoms. Of 24 plant and 19 vertebrate species experimentally inoculated with EBOV, only bats became infected. The bats displayed no clinical signs of disease, which is considered evidence that these bats are a reservoir species of EBOV. In a 2002–2003 survey of 1,030 animals including 679 bats from **Gabon and the Republic of the Congo**, 13 fruit bats were found to contain EBOV RNA. Antibodies against Zaire and Reston viruses have been found in fruit bats in **Bangladesh**, suggesting that these bats are also potential hosts of the virus and that the filoviruses are present in Asia. Filoviruses are another virus in the same family as the Ebola as well as Marburg viruses.

Between 1976 and 1998, in 30,000 mammals, birds, reptiles, amphibians and arthropods sampled from regions of EBOV outbreaks, no Ebola virus was detected apart from some genetic traces found in six rodents (belonging to the species *Mus setulosus* and *Praomys*) and one shrew (*Sylvisorex ollula*) collected from the Central African Republic. However, further research efforts have not confirmed rodents as a reservoir. Traces of EBOV were detected in the carcasses of gorillas and chimpanzees during outbreaks in 2001 and 2003, which later became the source of human infections. However, the high rates of death in these species resulting from EBOV infection make it unlikely that these species represent a natural reservoir for the virus.

Chapter 3: Virology

Ebolaviruses contain single-stranded, non-infectious RNA genomes. *Ebolavirus* genomes are approximately 19 kilobase pairs long and contain seven genes in the order 3'-UTR-*NP-VP35-VP40-GP-VP30-VP24-L*-5'-UTR. The genomes of the five different ebolaviruses (BDBV, EBOV, RESTV, SUDV and TAFV) differ in sequence and the number and location of gene overlaps. As all filoviruses, ebolavirions are filamentous particles that may appear in the shape of a shepherd's crook, of a "U" or of a "6," and they may be coiled, toroid or branched. In general, ebolavirions are 80 nanometers (nm) in width and may be as long as 14,000 nm.

Their life cycle begins with a virion attaching to specific cell-surface receptors, followed by fusion of the virion envelope with cellular membranes and the concomitant release of the virus nucleocapsid into the cytosol. The *Ebolavirus* structural glycoprotein (known as GP1,2) is responsible for the virus' ability to bind to and infect targeted cells. The viral RNA polymerase, encoded by the *L* gene, partially uncoats the nucleocapsid and transcribes the genes into positive-strand mRNAs, which are then translated into structural and nonstructural proteins. The most abundant protein produced is the nucleoprotein, whose concentration in the host cell determines when L switches from gene transcription to genome replication. Replication of the viral genome results in full-length, positive-strand antigenomes that are, in turn, transcribed into genome copies of negative-strand virus progeny.

Newly synthesized structural proteins and genomes self-assemble and accumulate near the inside of the cell membrane. Virions bud off from the cell, gaining their envelopes from the cellular membrane from which they bud from. The mature progeny particles then infect other cells to repeat the cycle. The genetics of the Ebola virus are difficult to study because of EBOV's virulent characteristics.

Similar to other filoviridae, EBOV replicates very efficiently in many cells, producing large amounts of virus in monocytes, macrophages, dendritic cells and other cells. Replication of the virus in monocytes triggers the release of high levels of inflammatory chemical signals.

EBOV is thought to infect humans through contact with mucous membranes or through skin breaks.

Once infected, endothelial cells (cells lining the inside of blood vessels), liver cells, and several types of immune cells such as macrophages, monocytes, and dendritic cells are the main targets of infection. Following infection with the virus, the immune cells carry the virus to nearby lymph nodes where further reproduction of the virus takes place. From there, the virus can enter the bloodstream and lymphatic system and spread throughout the body.
Macrophages are the first cells infected with the virus, and this infection results in programmed cell death. Other types of white blood cells, such as lymphocytes, also undergo programmed cell death leading to an abnormally low concentration of lymphocytes in the blood. This contributes to the weakened immune response seen in those infected with EBOV.

Filoviral infection also interferes with proper functioning of the innate immune system. EBOV proteins blunt the human immune system's response to viral infections by interfering with the cells' ability to produce and respond to interferon proteins such as interferon-alpha, interferon-beta, and interferon gamma.

The VP24 and VP35 structural proteins of EBOV play a key role in this interference. When a cell is infected with EBOV, receptors located in the cell's cytosol (such as RIG-I and MDA5) or outside of the cytosol (such as Toll-like receptor 3 (TLR3), TLR7, TLR8 and TLR9), recognize infectious molecules associated with the virus. On TLR activation, proteins including interferon regulatory factor 3 and interferon regulatory factor 7 trigger a signaling cascade that leads to the expression of type 1 interferons. The type 1 interferons are then released and bind to the IFNAR1 and IFNAR2 receptors expressed on the surface of a neighboring cell. Once interferon has bound to its receptors on the neighboring cell, the signaling proteins STAT1 and STAT2 are activated and move to the cell's nucleus. This triggers the expression of interferon-stimulated genes, which code for proteins with antiviral properties. EBOV's V24 protein blocks the production of these antiviral proteins by preventing the STAT1 signaling protein in the neighboring cell from entering the nucleus. The VP35 protein directly inhibits the production of interferon-beta. By inhibiting these immune responses, EBOV may quickly spread throughout the body.

Endothelial cells may be infected within 3 days after exposure to the virus. The breakdown of endothelial cells leading to vascular injury can be attributed to EBOV glycoproteins. The widespread hemorrhage that occurs in affected people causes edema and hypovolemic shock.

After infection, a secreted glycoprotein, small soluble glycoprotein (sGP) (or Ebola virus glycoprotein [GP]), is synthesized. EBOV replication overwhelms protein synthesis of infected cells and the host immune defenses. The GP forms a trimeric complex, which tethers the virus to the endothelial cells. The sGP forms a dimeric protein that interferes with the signaling of neutrophils, another type of white blood cell, which enables the virus to evade the immune system by inhibiting early steps of neutrophil activation.

The presence of viral particles and the cell damage resulting from viruses budding out of the cell causes the release of chemical signals (such as TNF-α, IL-6 and IL-8), which are molecular signals for fever and inflammation. The damage to human cells, caused by infection of the endothelial cells, decreases the integrity of blood vessels. This loss of vascular integrity increases with the synthesis of GP, which reduces the availability of specific integrins responsible for cell adhesion to the intercellular structure and causes damage to the liver, leading to improper clotting.

Chapter 4: Epidemiology and Outbreaks

The disease typically occurs in outbreaks in tropical regions of Sub-Saharan Africa. From 1976 (when it was first identified) through 2013, the World Health Organization reported 1,716 confirmed cases.

The largest outbreak to date is the ongoing 2014 West Africa Ebola virus outbreak, which is affecting Guinea, Sierra Leone, Liberia and Nigeria. As of November 2014, 14,413 suspected cases and 5,504 deaths had been reported.

The first known outbreak of EBOV was identified only after the fact, occurring between June and November 1976 in Nzara, South Sudan, (then part of Sudan) and was caused by Sudan virus (SUDV). The Sudan outbreak infected 284 people and killed 151. The first identifiable case in Sudan occurred on 27 June in a storekeeper in a cotton factory in Nzara, who was hospitalized on 30 June and died on 6 July. While the WHO medical staff involved in the Sudan outbreak were aware that they were dealing with an unknown disease, the actual "positive identification" process and the naming of the virus did not occur until some months later in the Democratic Republic of the Congo.

On 26 August 1976, a second outbreak of EBOV began in Yambuku, a small rural village in Mongala District in northern Zaire (now known as the Democratic Republic of the Congo). This outbreak was caused by EBOV, formerly designated Zaire ebolavirus, which is a different member of the genus Ebolavirus than in the first Sudan outbreak. The first person infected with the disease was village school headmaster Mabalo Lokela, who began displaying symptoms on 26 August 1976. Lokela had returned from a trip to Northern Zaire near the Central African Republic border, having visited the Ebola River between 12 and 22 August. He was originally believed to have malaria and was given quinine. However, his symptoms continued to worsen, and he was admitted to Yambuku Mission Hospital on 5 September. Lokela died on 8 September, 14 days after he began displaying symptoms.

Soon after Lokela's death, others who had been in contact with him also died, and people in the village of Yambuku began to panic. This led the country's Minister of Health along with Zaire President Mobutu

Sese Seko to declare the entire region, including Yambuku and the country's capital, Kinshasa, a quarantine zone. No one was permitted to enter or leave the area, with roads, waterways, and airfields were placed under martial law. Schools, businesses and social organizations were closed.

Researchers from the CDC, including Peter Piot, co-discoverer of Ebola, later arrived to assess the effects of the outbreak, observing that "the whole region was in panic."

Dr. Piot concluded that the Belgian nuns had inadvertently started the epidemic by giving unnecessary vitamin injections to pregnant women, without sterilizing the syringes and needles. The outbreak lasted 26 days, with the quarantine lasting 2 weeks. Among the reasons that researchers speculated caused the disease to disappear, were the precautions taken by locals, the quarantine of the area, and discontinuing the injections.

During this outbreak, Dr. Ngoy Mushola recorded the first clinical description of EEBOV in Yambuku, where he wrote the following in his daily log: "The illness is characterized with a high temperature of about 39 °C (102 °F), hematemesis, diarrhea with blood, retrosternal abdominal pain, prostration with "heavy" articulations, and rapid evolution death after a mean of 3 days."

The virus responsible for the initial outbreak, first thought to be Marburg virus, was later identified as a new type of virus related to Marburg viruses. Virus strain samples isolated from both outbreaks were named as the "Ebola virus" after the Ebola River, located near the originally identified viral outbreak site in Zaire. At the time, reports conflicted about who initially coined the name: Either Karl Johnson of the American CDC team or Belgian researchers, headed by Peter Piot. Subsequently a number of other cases were reported, almost all centered on the Yambuku mission hospital or having close contact with another case. 318 cases and 280 deaths occurred in Zaire. Although it was assumed that the two outbreaks were connected, scientists later realized that they were caused by two distinct ebolaviruses, SUDV and EBOV. The Zaire outbreak was contained with the help of the World Health Organization and transport from the Congolese air force, by quarantining villagers, sterilizing medical equipment, and providing protective clothing.

The second major outbreak occurred in Zaire (now the Democratic Republic of the Congo) in 1995, affecting 315 and killing 254.

In 2000, Uganda had an outbreak affecting 425 and killing 224; in this case the Sudan virus was found to be the Ebola species responsible for the outbreak.

In 2003 there was an outbreak in the Republic of the Congo that affected 143 and killed 128, a death rate of 90 percent, the highest death rate of a genus Ebolavirus outbreak to date.

In 2004 a Russian scientist died from Ebola after sticking herself with an infected needle.

Between April and August 2007, a fever epidemic in a four-village region of the Democratic Republic of the Congo was confirmed in September to have cases of Ebola. Many people who attended the recent funeral of a local village chief died. The 2007 outbreak eventually affected 264 individuals and resulted in the deaths of 187.

On 30 November 2007, the Uganda Ministry of Health confirmed an outbreak of Ebola in the Bundibugyo District in Western Uganda.

After confirmation of samples tested by the United States National Reference Laboratories and the Centers for Disease Control, the World Health Organization confirmed the presence of a new species of genus Ebolavirus, which was tentatively named Bundibugyo. The WHO reported 149 cases of this new strain and 37 of those led to deaths.

The WHO confirmed two small outbreaks in Uganda in 2012. The first outbreak affected 7 people and resulted in the death of 4 and the second affected 24, resulting in the death of 17. The Sudan variant was responsible for both outbreaks.

On 17 August 2012, the Ministry of Health of the Democratic Republic of the Congo reported an outbreak of the Ebola-Bundibugyo variant in the eastern region. Other than its discovery in 2007, this was the only time that this variant has been identified as responsible for an outbreak. The WHO revealed that the virus had sickened 57 people and claimed 29 lives. The probable cause of the outbreak was tainted bush meat hunted by local villagers around the towns of Isiro and Viadana.

In March 2014, the World Health Organization (WHO) reported a major Ebola outbreak in Guinea, a western African nation. Researchers traced the outbreak to a two-year old child who died December 2013. The disease then rapidly spread to the neighboring countries of Liberia and Sierra Leone. It is the largest Ebola outbreak ever documented, and the first recorded in the region.

On 8 August 2014, the WHO declared the epidemic to be an international public health emergency. Urging the world to offer aid to the affected regions, the Director-General said, "Countries affected to date simply do not have the capacity to manage an outbreak of this size and complexity on their own. I urge the international community to provide this support on the most urgent basis possible."

By mid-August 2014, Doctors Without Borders reported the situation in Liberia's capital Monrovia as "catastrophic" and "deteriorating daily". They reported that fears of Ebola among staff members and patients had shut down much of the city's health system, leaving many people without treatment for other conditions. By late August 2014, the disease had spread to Nigeria, and one case was reported in Senegal. On 30 September 2014, the first confirmed case of Ebola in the United States was diagnosed. The patient died 8 days later.

Aside from the human cost, the outbreak has severely eroded the economies of the affected countries. A Financial Times report suggested the economic impact of the outbreak could kill more people than the virus itself. As of 23 September, in the three hardest hit countries, Liberia, Sierra Leone and Guinea, only 893 treatment beds were available even though the current need was 2122 beds.

In a 26 September statement, the WHO said, "The Ebola epidemic ravaging parts of West Africa is the most severe acute public health emergency seen in modern times. Never before in recorded history has a biosafety level four pathogen infected so many people so quickly, over such a broad geographical area, for so long." The WHO reported that by 25 August more than 216 health-care workers were among the dead, partly due to the lack of equipment and long hours.

On 23 October, the Malian government confirmed its first case. In response, UNMEER, in cooperation with the Logistics Cluster, air-lifted 1,050 kg of personal protective equipment (PPE) and body bags from Monrovia to Mali. As of 11 November 2014, 14,413 suspected cases and 5,504 deaths had been reported; however, the WHO has said that these numbers may be vastly underestimated.

An outbreak in Boende District in Equatorial Province was stopped effectively with flexible organization and funding, as well as social mobilization led by UNICEF advising action people could use. The DRC outbreak was from a local Ebola strain and not the one from West Africa (WHO).

As of 15 October 2014, there have been 17 cases of Ebola treated outside of Africa, four of whom have died. In early October, Teresa Romero, a 44-year-old Spanish nurse, contracted Ebola after caring for a priest who had been repatriated from West Africa. This was the first transmission of the virus to occur outside of Africa. On 20 October, it was announced that Teresa Romero had tested negative for the Ebola virus, suggesting that she may have recovered from Ebola infection.

On 19 September, Eric Duncan flew from his native Liberia to Texas; 5 days later he began showing symptoms and visited a hospital, but was sent home. His condition worsened and he returned to the hospital on 28 September, where he died on 8 October. Health officials confirmed a diagnosis of Ebola on 30 September—the first case in the United States. On 12 October, the CDC confirmed that a nurse in Texas who had treated Duncan was found to be positive for the Ebola virus, the first known case of the disease to be contracted in the United States. On 15 October, a second Texas health-care worker who had treated Duncan was confirmed to have the virus. They have both recovered.

On 23 October, a doctor in New York City, who returned to the United States from Guinea after working with Doctors Without Borders, tested positive for Ebola. His case is unrelated to the Texas cases.

Chapter 5: The Reston ebolavirus

In late 1989, Hazelton Research Products' Reston Quarantine Unit in Reston, Virginia, suffered an outbreak of fatal illness amongst certain lab monkeys. This lab outbreak was initially diagnosed as simian hemorrhagic fever virus (SHFV), and occurred amongst a shipment of crab-eating macaque monkeys imported from the Philippines. Hazelton's veterinary pathologist sent tissue samples from dead animals to the United States Army Medical Research Institute of Infectious Diseases (USAMRIID) at Fort Detrick, Maryland, where an ELISA test indicated the antibodies present in the tissue were a response to Ebola virus and not SHFV. An electron microscopist from USAMRIID discovered filoviruses similar in appearance to Ebola in the tissue samples sent from Hazelton Research Products' Reston Quarantine Unit.

A US Army team headquartered at USAMRIID euthanized the surviving monkeys, and brought all the monkeys to Fort Detrick for study by the Army's veterinary pathologists and virologists, and eventual disposal under safe conditions. Blood samples were taken from 178 animal handlers during the incident.

All of those six animal handlers eventually seroconverted, including one who had cut himself with a bloody scalpel. Despite its status as a Level-4 organism and its apparent pathogenicity in monkeys, when the handlers did not become ill, the CDC concluded that the virus had a very low pathogenicity to humans.

The Philippines and the United States had no previous cases of Ebola infection, and upon further isolation, researchers concluded it was another strain of Ebola, or a new filovirus of Asian origin, which they named *Reston ebolavirus* (RESTV) after the location of the incident. Reston virus (RESTV) can be transmitted to pigs. Since the initial outbreak it has since been found in nonhuman primates in Pennsylvania, Texas, and Italy, where the virus had infected pigs. According to the WHO, routine cleaning and disinfection of pig (or monkey) farms with sodium hypochlorite or detergents should be effective in inactivating the *Reston ebolavirus*. Pigs that have been infected with RESTV tend to show symptoms of the disease.

Chapter 6: Research and Vaccines

A number of experimental treatments are being studied. In the United States, the Food and Drug Administration (FDA)'s animal efficacy rule is being used to demonstrate reasonable safety to obtain permission to treat people who are infected with Ebola. It is being used because the normal path for testing drugs is not possible for diseases caused by dangerous pathogens or toxins. Experimental drugs are made available for use with the approval of regulatory agencies under named patient programs, known in the US as "expanded access".

On 12 August 2014 the WHO released a statement that the use of not yet proven treatments is ethical in certain situations in an effort to treat or prevent the disease.

A number of antiviral medications are being studied.

- **Favipiravir**, approved in Japan for stockpiling against influenza pandemics, appears to be useful in a mouse model of Ebola. On 4 October 2014, it was reported that a French nun who contracted Ebola while volunteering in Liberia was cured with a Favipiravir treatment.

- **BCX4430** is a broad-spectrum small molecule antiviral drug developed by BioCryst Pharmaceuticals and undergoing animal testing as a potential human treatment for Ebola by USAMRIID. The drug has been approved to progress to Phase 1 trials, expected late in 2014.

- **Brincidofovir** is a broad-spectrum antiviral drug. Its manufacturer has been granted FDA approval to proceed with a trial to test its safety and efficacy in Ebola patients. It has been used to treat the first patient diagnosed with Ebola in the USA, after he had recently returned from Liberia.

- **Lamivudine**, usually used to treat HIV/AIDS, was reported in September 2014 to have been used successfully to treat 13 out of 15 Ebola-infected patients by a doctor in Liberia, as part of a combination therapy also involving intravenous fluids and antibiotics to combat opportunistic bacterial infection of Ebola-compromised internal organs. Western virologists have however expressed caution about the results, due to the small number of patients treated and confounding factors present. Researchers at the National Institute of Health (NIH) stated that lamivudine had so far failed to demonstrate anti-Ebola activity in preliminary *in vitro* tests, but that they would continue to test it under different conditions and would progress it to trials if even slight evidence for efficacy is found.

- **JK-05** is developed by the Chinese company Sihuan Pharmaceutical along with the Chinese Academy of Military Medical Sciences. It is reportedly being fast tracked through human trials for Ebola treatment after successful tests in mice.
- Lack of available treatment options has spurred research into a number of other possible antivirals targeted against Ebola, including natural products such as **scytovirin** and **griffithsin**, as well as synthetic drugs including **DZNep**, **FGI-103, FGI-104, FGI-106, dUY11** and **LJ-001**, and other newer agents.

Other promising treatments rely on antisense technology. Both small interfering RNAs (siRNAs) and phosphorodiamidate morpholino oligomers (PMOs) targeting EBOV RNA polymerase L protein may prevent disease in nonhuman primates. TKM-Ebola is a small interfering RNA compound, currently being tested in a Phase I clinical trial in humans.

Sarepta Therapeutics has completed a Phase I clinical trial with its PMO protecting up to 80-100 percent of the nonhuman primates tested.

Antibodies

ZMapp is a monoclonal antibody vaccine. The limited supply of the drug has been used to treat a small number of individuals infected with the Ebola virus. Although some individuals have recovered, the outcome is not considered statistically significant. ZMapp has proved effective in a trial involving rhesus macaque monkeys.

The Bill & Melinda Gates Foundation has donated $150,000 to help Amgen increase its production, and the U.S. Department of Health and Human Services has asked a number of centers to also increase production.

There was no confirmation or proof that the ZMapp drug was a factor in the recovery of two American Ebola patients, however; furthermore, a Spanish priest with Ebola had taken ZMapp and died afterward.

Researchers in Thailand claim to have developed an antibody-based treatment for Ebola using synthesized fragments of the virus. It has not been tested against Ebola itself. Scientists from the WHO and NIH have offered to test the treatment against live Ebola virus, but there is still a great deal of development needed before human trials.

Other

Two selective estrogen receptor modulators usually used to treat infertility and breast cancer (clomiphene and toremifene) have been found to inhibit the progress of Ebola virus *in vitro* as well as in infected mice. Ninety percent of the mice treated with clomiphene and 50 percent of those treated with toremifene survived the tests. The study authors conclude that given their oral availability and history of human use, these drugs would be candidates for treating Ebola virus infection in remote geographical locations, either on their own or together with other antiviral drugs.

A 2014 study found that three ion channel blockers used in the treatment of heart arrhythmias, amiodarone, dronedarone and verapamil, block the entry of Ebola virus into cells *in vitro*.

Blood products

The WHO has stated that transfusion of whole blood or purified serum from Ebola survivors is the therapy with the greatest potential to be implemented immediately, although there is little information as to its efficacy.

In September 2014, WHO issued an interim guideline for this therapy. The blood serum from those who have survived an infection is currently being studied to see if it is an effective treatment. During a meeting arranged by WHO, this research was deemed to be a top priority. Seven of eight people with Ebola survived after receiving a transfusion of blood donated by individuals who had previously survived the infection in an 1999 outbreak in the Democratic Republic of the Congo. This treatment, however, was started late in the disease meaning they may have already been recovering on their own and the rest of their care was better than usual.

However, this potential treatment remains controversial. Intravenous antibodies appear to be protective in nonhuman primates who have been exposed to large doses of Ebola. The WHO has approved the use of convalescent serum and whole blood products to treat people with Ebola.

Vaccine

Many Ebola vaccine candidates had been developed in the decade prior to 2014, but as of October 2014, none had yet been approved by the United States Food and Drug Administration (FDA) for clinical use in humans. Several promising vaccine candidates have been shown to protect nonhuman primates (usually macaques) against lethal infection. These include replication-deficient adenovirus vectors, replication-competent vesicular stomatitis (VSV) and human parainfluenza (HPIV-3) vectors, and virus-like particle preparations.

Conventional trials to study efficacy by exposure of humans to the pathogen after immunization are obviously not feasible in this case. For such situations, the FDA has established the "animal rule" allowing licensure to be approved on the basis of animal model studies that replicate human disease, combined with evidence of safety and a potentially potent immune response (antibodies in the blood) from humans given the vaccine.

Phase I clinical trials involve the administration of the vaccine to healthy human subjects to evaluate the immune response, identify any side effects and determine the appropriate dosage.

In September 2014, two Phase I clinical trials began for the vaccine cAd3-EBO Z, which is based on an attenuated version of a chimpanzee adenovirus (cAd3) that has been genetically altered so that it is unable to replicate in humans. It was developed by NIAIDin collaboration with Okairos, now a division of GlaxoSmithKline. For the trial designated VRC 20, 20 volunteers were recruited by the NIAID in Bethesda, Maryland, while three dose-specific groups of 20 volunteers each were recruited for trial EBL01 by University of Oxford, U.K.

A replication-competent vaccine based on the vesicular stomatitis virus, called VSV-EBOV, was developed by the Canadian National Microbiology Laboratory and licensed to the small company NewLink Genetics. With the strong support of the U.S. Defense Threat Reduction Agency, it started Phase I clinical trials on healthy human subjects on 13 October 2014 at the Walter Reed Army Institute of Research in Silver Spring, Md.

Also in October 2014, the U.S. National Institute of Allergy and Infectious Diseases(NIAID) was recruiting healthy human volunteers for a "Phase 1 Randomized, Double-Blind, Placebo Controlled, Dose-Escalation Study to Evaluate the Safety and Immunogenicity of Prime-Boost VSV Ebola Vaccine in Healthy Adults".

On 20 October, the Public Health Agency of Canada began air shipment of 800 doses of the VSV-EBOV vaccine to the WHO in Geneva. This vaccine is intended to be used in Phase I clinical trials, to start in late October or early November. The WHO has recruited 250 volunteers ready to begin Phase I clinical trials in four locations: Switzerland, Germany, Gabon and Kenya. If the results of this and following trials show that the earlier results in nonhuman primates are replicable in humans, this vaccine could be deployed in areas such as West Africa and would be expected to require only a single dose. Also, its efficacy in protecting nonhuman primates when administered even after viral exposure has occurred may help protect health-care workers after a suspected exposure.

The Health Ministry of Russia also claims to have developed a vaccine called Triazoverin, which is said to be effective against both Ebola and Marburg filoviruses, and might be available for clinical trials in West Africa as soon as the start of 2015.

At the 8th Vaccine and ISV Conference in Philadelphia on 27–28 October 2014, Novavax Inc. reported the development in a "few weeks" of a glycoprotein (GP) nanoparticle Ebola virus (EBOV GP) vaccine using their proprietary recombinant technology. A recombinant protein is a protein whose code is carried by recombinant DNA. The vaccine is based on the newly published genetic sequence of the 2014 Guinea Ebola strain that is responsible for the current Ebola disease epidemic in West Africa. In "preclinical models", a useful immune response was induced, and was found to be enhanced ten to a hundred-fold by the company's "Matrix-M" immunologic adjuvant.

A study of the response of non-human primate to the vaccine had been initiated. Attractive features of such a vaccine could be no need for frozen storage, and the possibility of rapid scaling to manufacture of large dose quantities. Novavax has initiated a primate study and expects to initiate a Phase 1 clinical trial in December 2014.

Chapter 7: Initial 1976 Discovery

Below is an actual interview in July 2014, with the man who discovered the Ebola virus, back in 1976, Dr. Peter Piot, through Rob Brown, of the BBC World Service. It lends some insight how the Ebola virus was initially discovered, viewed at the time, and how it evolved into the deadly strain it is today throughout the current 2014, and on into 2015 epidemic.

In September 1976, a package containing a shiny, blue thermos flask arrived at the Institute of Tropical Medicine in Antwerp, Belgium.

Working in the lab that day was Peter Piot, a 27-year-old scientist and medical school graduate training as a clinical microbiologist.

"It was just a normal flask like any other you would use to keep coffee warm," recalls Dr. Piot, now Director of the London School of Hygiene and Tropical Medicine.

But this thermos wasn't carrying coffee - inside was an altogether different cargo. Nestled amongst a few melting ice cubes were vials of blood along with a note.

It was from a Belgian doctor based in what was then Zaire, now the Democratic Republic of Congo - his handwritten message explained that the blood was that of a nun, also from Belgium, who had fallen ill with a mysterious illness which he couldn't identify.

This unusual delivery had travelled all the way from Zaire's capital city Kinshasa, on a commercial flight, in one of the passengers' hand luggage.

"When we opened the thermos, we saw that one of the vials was broken and blood was mixing with the water from the melted ice," recalled Dr. Piot.

He and his colleagues were unaware just how dangerous that was. As the blood leaked into the icy water so too did a deadly unknown virus.

The samples were treated like numerous others the lab had tested before, but when the scientists placed some of the cells under an electron microscope they saw something they didn't expect.

"We saw a gigantic worm like structure - gigantic by viral standards," said Dr. Piot. "It's a very unusual shape for a virus, only one other virus looked like that and that was the Marburg virus."

The Marburg virus was first recognized in 1967 when 31 people became ill with hemorrhagic fever in the cities of Marburg and Frankfurt in Germany and in Belgrade, the capital of Yugoslavia. This Marburg outbreak was associated with laboratory staff who were working with infected monkeys imported from Uganda, in which seven people died.

Dr. Piot knew how serious the Marburg could be, but after consulting with experts around the world he received confirmation that what he was seeing under the microscope wasn't the Marburg virus. This was something else, something never seen before.

"It's hard to describe but the main emotion I had was one of real, incredible excitement," Dr. Piot explained.

"There was a feeling of being very privileged, that this was a moment of discovery," her further elaborated.

News had reached Antwerp that the nun, who was under the care of the doctor in Zaire, had died. The team also learned that many others were falling ill with this mysterious illness in a remote area in the north of the country and their symptoms included fever, diarrhea and vomiting followed by bleeding and eventually death.

Two weeks later Dr. Piot, who had never been to Africa before, was on a flight to Kinshasa.

"It was an overnight flight and I couldn't sleep. I was so excited about seeing Africa for the first time, about investigating this new virus and about stopping the epidemic." Dr. Piot recalled.

The journey didn't end in Kinshasa, and the team had to travel to the center of the outbreak, a village in the equatorial rainforest, about 1,000km (620 miles) further north.

"The personal physician of President Mobutu, the leader of Zaire at that time, arranged a C-130 transport aircraft for us," Dr. Piot remembered.

They loaded a Landrover, fuel and all the equipment they needed on to the plane.

When the C-130 landed in Bumba, a river port situated on the northernmost point of the Congo River, the fear surrounding the mysterious disease was tangible. Even the pilots didn't want to hang around for long and they actually kept the airplane's engines running as the team unloaded their kit.

"As they left they shouted 'Adieu,'" Dr. Piot remembered with great accuracy.

"In French, people say 'Au Revoir' to say 'See you again', but when they say 'Adieu' - well, that's like saying, 'We'll never see you again,'" Dr. Piot further explained.

Standing on the tarmac watching the plane leave, facing a deadly unknown virus in an unfamiliar place, some people might have regretted the decision to go there.

"I wasn't scared. The excitement of discovery and wanting to stop the epidemic was driving everything.

We heard far more people were dying from the disease than we originally thought and we wanted to get to work," Dr. Piot continued.

The curiosity and sense of adventure that brought Dr. Piot to this point had been ignited many years earlier when he was a young boy growing up in a small rural village in the Flanders region of Belgium.

A museum near Dr. Piot's home was dedicated to a local saint who worked with leprosy patients, and it was here that he received his first glimpse into the world of disease and microbiology.

"I decided one day to cycle to the museum. The old pictures I saw there of those suffering from leprosy fascinated me," he said.

"That sparked my interest in medicine - it gave me a thirst for scientific knowledge, a desire to help people and I hoped it would give me a passport to the world," a young Peter Piot recalled from his boyhood.

It did give Dr. Piot a passport to the world. The team's final destination was the village of Yambuku - about 120km (75 miles) from Bumba, where the plane had left them.

Yambuku was home to an old Catholic mission, and it had a hospital and a school run by a priest and nuns, whereby all of them from Belgium.

"The area was beautiful. The mission was surrounded by lush rainforest and the earth was red - the nature was incredibly rich but the people were so poor," Dr. Piot remembered.

"Joseph Conrad called that place 'The Heart of Darkness', but I thought there was a lot of light there," he further explained.

The beauty of Yambuku belied the horror that was unfolding for the people that lived there.

When Dr. Piot arrived, the first people he met were a group of nuns and a priest who had retreated to a guesthouse and established their own cordon sanitaire: A barrier used to prevent the spread of disease.

There was a sign on the cord, written in the local Lingala language that read, "Please stop, anybody who crosses here may die."

"They had already lost four of their colleagues to the disease," Dr Piot recalled.

"They were praying and waiting for death," he continued.

Dr. Piot jumped over the cordon and told them that the team would help them and stop the epidemic.

"When you are 27, you have all this confidence," he said.

The nuns told the newly arrived scientists what had happened, they spoke about their colleagues and those in the village who had died and how they tried to help as best they could.

The priority was to stop the epidemic, but first the team needed to find out how this virus was moving from person to person: By air, in food, by direct contact or spread by insects?

"We had to start asking questions. It was really like a detective story," Dr. Piot recalled like it was yesterday.

These were the three questions they asked:

• How did the epidemic evolve? Knowing when each person caught the virus gave clues to what kind of infection this was, and from here the story of the virus began to emerge.

• Where did the infected people come from? The team visited all the surrounding villages and mapped out the number of infections. It was clear that the outbreak was closely related to areas served by the local hospital.

• Who gets infected? The team found that more women than men caught the disease and particularly women between 18 and 30 years old. It turned out that many of the women in this age group were pregnant and many had attended an antenatal clinic at the hospital.

The mystery of the virus was beginning to unravel.

The team then discovered that the women who attended the antenatal clinic all received a routine injection. Each morning, just five syringes would be distributed, the needles would be reused and so the virus was spread between the patients.

"That's how we began to figure it out," recalled Dr. Piot.

"You do it by talking, looking at the statistics and using logical deduction," he explained.

The team also noticed that people were getting ill after attending funerals. When someone died from Ebola, the body is full of the virus - any direct contact, such as washing or preparation of the deceased without protection can be a serious risk.

The next step was to stop the transmission of the virus.

"We systematically went from village to village and if someone was ill they would be put into quarantine," Dr. Piot further explained.

"We would also quarantine anyone in direct contact with those infected and we would ensure everyone knew how to correctly bury those who had died from the virus," he reported.

The closure of the hospital, the use of quarantine and making sure the community had all the necessary information eventually brought an end to the epidemic, but nearly 300 people died.

Dr. Piot and his colleagues had learned a lot about the virus during their three months in Yambuku, but it still lacked a name.

"We didn't want to name it after the village, Yambuku, because it's so stigmatizing. You don't want to be associated with that," Dr. Piot went into detail.

The team decided to name the virus after a river. They had a map of Zaire, although not a very detailed one, and the closest river they could see was the Ebola River. From that point on, the virus that arrived in a flask in Antwerp all those months earlier would be known as the Ebola virus.

In February 2014, Dr. Piot returned to Yambuku for only the second time since 1976, to mark his 65th birthday. He met Sukato Mandzomba, one of the few who caught the virus in 1976 and survived.

"It was fantastic to meet him again, it was a very moving moment," recalls Dr. Piot.

Back then, Mandzomba was a nurse in the local hospital and could speak French so the pair had managed to build up a rapport.

"He's still living in Yambuku and still working in the hospital - he's now running the lab there and it's impeccable. I was really impressed," Dr. Piot further recalled.

It's 38 years since that initial outbreak and the world is now experiencing its worst Ebola epidemic ever. So far more than 600 people have died in the West African countries of Guinea, Liberia and Sierra Leone. The current situation has been called unprecedented, the spread of the disease across three countries making it more complicated to deal with than ever before.

In the absence of any vaccine or cure, the advice for this outbreak is much the same as it was in the 1970s.

"Soap, gloves, isolating patients, not reusing needles and quarantining the contacts of those who are ill - in theory it should be very easy to contain Ebola," Dr. Piot explained.

In practice though, other factors can make fighting an Ebola outbreak a difficult task. People who become ill and their families may be stigmatized by the community, resulting in a reluctance to come forward for help. Cultural beliefs lead some to think the disease is caused by witchcraft, while others are hostile towards health workers.

"We shouldn't forget that this is a disease of poverty, of dysfunctional health systems - and of distrust," Dr. Piot points out.

For this reason, information, communication and involvement of community leaders are as important as the classical medical approach, he argues.

Ebola changed Dr. Piot's life: Following the discovery of the virus, he went on to research the Aids epidemic in Africa and became the founding executive director of the UNAIDS organization.

"It led me to do things I thought only happened in books. It gave me a mission in life to work on health in developing countries," he said.

"It was not only the discovery of a virus but also of myself," Dr. Piot concluded, at least for this interview.

Chapter 8: The 2014 Strain

Below is another interview, in July 2014 as well, with Dr. Peter Piot, through CNN's Christine Amanpour, this one specifically discussing the 2014 epidemic.

The scientist who discovered the Ebola virus said that a current outbreak of the deadly bug in West Africa, in which 467 people have died, is "unprecedented."

"One, this is the first time in West Africa that we have such an outbreak," Dr. Peter Piot told CNN's Christiane Amanpour.

"Secondly, it is the first time that three countries are involved. And thirdly it's the first time that we have outbreaks in capitals, in capital cities," he elaborated.

Doctors Without Borders warns that the outbreak in Guinea, Sierra Leone, and Liberia is now "out of control." The number of cases is still spiking since it was first observed around the beginning of this year.

"With this strain of Ebola, you've got like a ninety percent chance of dying. That's spectacular by any standard – one of the most lethal viruses that exist," Dr. Piot stated.

And the way victims die is far from pleasant.

"Ebola virus infection starts with something that looks like the flu – headache, fever, maybe diarrhoea. But then you can develop very fast bleeding that's uncontrollable, and that's how people die," Dr. Piot described the terror.

There is no cure for Ebola, but in theory the disease should be easy to fight, Dr. Piot said.

"You need really close contact to become infected. So just being on the bus with someone with Ebola, that's not a problem," He gave insight.

"Simple hygienic measures like washing with soap and water, not re-using syringes, and avoiding contact with infected corpses are sufficient to stop spread of the disease," Dr. Piot said.

"This is an epidemic of dysfunctional health systems," he continued to elaborate.

"Fear of the virus, and the lack of trust in government, in the health system, is as bad as the actual virus," he observed.

"What happens is that a person is infected, is hospitalized, infects other patients and particularly health care workers," Dr. Piot sid.

"They're buried somewhere; around that funeral, people are infected when they touch the body, and so on. And then they get ill, and then they go somewhere else, and then they go to relatives in town, maybe because they hope to have better health care. That's how it spreads," he informed.

Dr. Piot was a young researcher at the Institute of Tropical Medicine in Antwerp in the 1970s when they got a blood sample from a Belgian nun who had died in Zaire (now the Democratic Republic of Congo).

"The clinical diagnosis was yellow fever – and we were equipped to isolate the yellow fever virus. But then something completely different came out of it, and under the microscope it looked like – kind of a more like a wormlike structure," he recalls with accurate detail.

It was, he said, "a very big virus. At the time, there was only one other virus known to be of that shape: The Marburg Virus," he explained.

They shipped the sample off to the Centers for Disease Control in the U.S., who confirmed that it was indeed something completely new.

But forty years later, much is still unknown about Ebola.

"We are not one hundred percent sure where this virus comes from: Probably from some bat," he said.

"A lot of research has been done, but not enough. And as long as we don't know where exactly this virus hides outside epidemics, then we can't map where the risks are," he furthered.

The current outbreak is "already a mega-crisis," Dr. Piot said.

"For me, this is a reason for a state of emergency, you know, in these countries," he said.

"You need a combination of nearly military type of control measures – isolation, quarantine of those who are the diseased – but also their relatives, to make sure that they're not spreading the infection."

"And, secondly, community mobilization. Information can save lives here," Dr. Piot reported.

That was brought into sharp focus on Wednesday, when the Red Cross said that it would suspend operations in Guinea after some of its workers were threatened by men with knives.

"The fear of the virus and running away from health services, that contributes to perpetuate the spread of Ebola virus," Dr. Piot observed.

Ebola is a virus, yet comes in many different strains.

The fear is that it might begin to behave like a bacteria, becoming immune to antibiotics, and UK Prime Minister David Cameron announced a new initiative Wednesday to try to tackle the problem.

"This is one of the biggest challenges of the future, and that is the possibility that some of the very banal bacterial infections will be untreatable," Dr. Piot took a different angle.

"And then major surgery will become untreatable, people admitted to hospital will die from these infections. We already have some strains like that for tuberculosis," he noticed.

The solution is a "more disciplined" prescription of antibiotics, curtailing the amount of antibiotics given to animals (which ends up in our food), and the development on new antibiotics.

"I think, now, because of the greater awareness, there is hope. But we're running out of time," Dr. Piot concluded this interview on the 2014 strain of the Ebola virus outbreak.

Chapter 9: Current Death Count & Outlook

The number of Ebola cases so far this year: 9,936. How many people have been killed by Ebola: 4,877. These are the official figures put out by the World Health Organization, widely regarded as the authority on the Ebola outbreak in West Africa. Those statistics have been widely circulated, reported, tweeted and retweeted, but the number of deaths related to Ebola is based largely on speculation, not concrete evidence, according to a spokesman for WHO.

"We don't really know how many deaths there have been, because there are a lot of people who have died alone ... or out in the bush," WHO spokesman Dan Epstein said.

Those deaths go unreported, Epstein noted, so the agency uses statistical models to account for what they estimate is the number of unreported cases. Health workers on the ground in Guinea,

Liberia and Sierra Leone often rely on anecdotal evidence to estimate the number of Ebola deaths in a particular village or town. During field studies, locals might tell disease trackers that their village has had 20 deaths from Ebola, but "no one is verifying it," Epstein said.

WHO most recently said it believes the actual Ebola death toll is about three times higher than the roughly 4,900 confirmed or suspected cases. That would bring the number of people killed in the current Ebola outbreak to about 15,000. The startling figure is just one of many from WHO since the outbreak in West African began in March. Earlier this month, the organization said the weekly infection rate of Ebola could reach 10,000 new cases by December, which would be a sharp spike from the current infection rate of about 1,000 cases per week. In early September, WHO said as many as 20,000 people could become infected with the virus by the time the outbreak is over.

How the death rate can escalate from about 1,000 new Ebola cases per week to a suspected 10,000 new Ebola cases per week within two months is still unclear. For WHO assistant director-general Bruce Aylward, who is leading the research on the outbreak, estimating the number of future Ebola cases is "both art and science," Epstein said. "Part of it was based on modeling and part of it was a guess," he said. Estimating the number of unreported Ebola cases involves using a correction factor of 2.0, according to Science Mag. That means WHO doubles the number of confirmed cases to roughly approximate its figures.

WHO gets its Ebola death numbers from three main sources: Clinics and medical centers; laboratories doing Ebola tests; and burial teams. In any outbreak, there are always some cases that slip through the cracks. These patients are not counted in official disease tallies, leaving health researchers with the difficult task of figuring out just how many cases are not reported and factoring that into their "official" numbers. The health agency said the discrepancy comes down to underreporting.

Keeping track of Ebola in West Africa is particularly difficult, health experts say, given that so many patients either never visit a health facility or are turned away because of overcrowding. Also, health workers have had to combat widespread distrust of health care workers among many victims of the disease who feel stigmatized and may not come forward when they are feeling symptoms. They might die quietly in their homes, unknown to WHO. "In the African nations currently experiencing the outbreak, patients don't trust their local health care systems – sometimes with good reason," Reuters reported in July. Locals often turn to traditional healers, meaning their official diagnoses are never recorded.

WHO has acknowledged that its numbers are "vastly" underestimated. But without properly verifying the number of reported deaths, there is still "a tremendous amount of uncertainty," WHO epidemiologist Christopher Dye stated in a recent Science Magazine article.

Postscript

What doesn't kill us makes us stronger.

~unknown

References (in order of use)

(1) http://en.wikipedia.org/wiki/Ebola_virus_disease

(2) http://web.orange.co.uk/article/news/ebola_outbreak_officials_confirm_14_dead

(3) http://www.bbc.com/news/magazine-28262541

(4) http://www.dispatch.com/content/stories/insight/2014/10/12/1-fatal-discovery.html

(5) http://www.spiegel.de/international/world/interview-with-peter-piot-discoverer-of-the-ebola-virus-a-993111.html

(6) http://amanpour.blogs.cnn.com/2014/07/02/scientist-who-discovered-ebola-this-is-unprecedented/

(7) http://www.dailymail.co.uk/sciencetech/article-2810005/Ebola-s-family-history-revealed-Scientists-discover-ancestors-killer-virus-23-MILLION-years-old-lead-new-vaccines.html

(8) http://www.businessinsider.com/peter-piot-tells-the-ebola-discovery-story-2014-8

(9) http://www.npr.org/blogs/goatsandsoda/2014/08/29/344257046/the-co-discoverer-of-ebola-never-imagined-an-outbreak-like-this

(10) http://www.reuters.com/article/2014/03/25/us-ebola-scientist-idUSBREA2O0VP20140325

(11) http://www.naturalnews.com/047223_ebola_viral_pandemic_catastrophe.html

(12) http://www.nytimes.com/2014/08/08/science/a-witness-to-ebolas-discovery.html?_r=0

(13) http://www.ibtimes.com/ebola-death-toll-2014-how-many-people-have-really-died-virus-1711477

(14) http://quoteperfect.net/h-g-wells-quotes/

(15) http://www.pinterest.com/pin/110408628334568562/

(16) http://www.great-quotes.com/quotes/author/H.+G./Wells

About the Author

Gwendolyn Olmsted, MBA, the former Washington DC microbiologist and research scientist, who conducted graduate level research at major universities as well as United States federal governmental facilities, moved to Florida in 2010. She has a Bachelor's and Master's degree in Business Finance, as well as a second Bachelor's and Master's degree in Environmental Sciences & Sustainability from both private and public universities, now writes stories from the Investigative True Crime, Romance, Biographical, Documentary, and Investigative Journalism genres.

www.ingramcontent.com/pod-product-compliance
Lightning Source LLC
Chambersburg PA
CBHW071758170526
45167CB00003B/1086